YOUR KNOWLEDGE HAS VALUE

- We will publish your bachelor's and
  master's thesis, essays and papers

- Your own eBook and book -
  sold worldwide in all relevant shops

- Earn money with each sale

Upload your text at www.GRIN.com
and publish for free

Bibliographic information published by the German National Library:

The German National Library lists this publication in the National Bibliography; detailed bibliographic data are available on the Internet at http://dnb.dnb.de .

Imprint:

Copyright © 2015 GRIN Verlag, Open Publishing GmbH
Print and binding: Books on Demand GmbH, Norderstedt Germany
ISBN: 978-3-668-05546-9

This book at GRIN:

http://www.grin.com/en/e-book/306464/challenges-and-difficulties-of-living-in-river-deltas-a-review-of-the

Masoud Kazem

Challenges and Difficulties of Living in River Deltas. A Review of the Major River Deltas in Asia and Africa

GRIN Publishing

LIVING IN RIVER DELTAS: CHALLENGES AND DIFFICULTIES

A REVIW OVER THE MAJOR RIVER DELTAS IN ASIA AND AFRICA

Masoud Kazem[1,2]

1-PhD. Student, Faculty of Society and Design, Bond University, Queensland Australia

2-Water Engineer, Tavan Ab consulting engineering co., Tehran, Iran

ABSTRACT

There is growing concerns owing the poor quality of life amongst the residents in the river deltas, particularly in the developing countries. In this paper, the current published researches have been investigated to classify the main outlined reasons of degradation of the life quality in the Asia and Africa's river deltas .The findings indicate that; although there is a common spectrum of similar reasons in both continents, but the importance sequences of the factors are dissimilar. In Asia, high population growth beside the extensive industrialization policies is threatening the river deltas. In contrast, in Africa, poverty and low level of adaptation capability have been mentioned as the main reasons exacerbating the impacts of climate change; a sequence of destructive factors that have reduced the life quality in the river deltas.

Keywords: River Delta, Life Quality, Industrialization, Poverty, Climate Change

Inhalt

Introduction

River deltas have been one of the most important regions in the evolution of the human civilization. The initial form of human settlement shaped in these regions because access to fresh water and fertile soil were the golden keys of the agriculture-based civilizations. Even after industrial revolution, delta regions keep their importance due to the fact that these places contained the highest level of wealth and prosperity. However, nowadays, a glance over the statistics and media reveals that poverty, overcrowded unhealthy districts and shanty towns have become an interpretable feature of many of river deltas, especially in the developing countries. Now it seems that the human kind faces an important question; "Nowadays, in what extent living in river deltas can secure life standards such as access to healthy food, fresh water, sanitation, job opportunities and other welfare factors?"

In order to answer this question, we should consider that the modern Homo-Sapience has changed the appearance of river deltas during less than one century. The unique features of our water containing planet that have been constant for thousands of years, now changed extensively. Almost all around the world, the dam construction policy has accelerated since the end of Second World War (Susser , 1985, Lowi, 1995 and Wang et al, 2013). Not only in Europe and United States; Asian large rivers (such as Yellow, Yangtze, Pearl, Chao Phraya, Indus, Krishna, Godavari, etc.) have been also affected by this policy. According to a comprehensive research, the sediment transportation in these rivers has been disrupted in a way that the total sediment load reduced by more than 80% over last 60 years (Liu and Lu, 2014). It means that the soil fertility has been reduced and now, more amounts of fertilizers are required. Consequently, the higher level of poisons and chemical substances (some of them with negative and some others with unknown effects on the environment) has been injected in the soil and water body of river deltas. The next consequences of this disaster have shaped a vicious circle threatening the survival of delta regions. If the negative effects of other human activities such as navigation, salt farms, industrial and urban disposals and fishery being considered, the seriousness of this tragedy would become more tangible. However, the human activities must not be considered as the only cause of these challenges. Consequently, answering to the mentioned question needs a comprehensive study on the long list of river deltas all around the world. As a first round assessment approach, a review over the current researches can reveal the main difficulties that residents in the delta regions are challenging with.

This article contains a review over the recent researches in various fields in order to provide a comprehensive image of the human made disasters in the major deltas in two different continents; Asia and Africa. Although there are common difficulties in the delta regions of the both continents, the main reason of the challenges may be slightly different. In the next paragraphs, the results of a series of current researches will be outlined in order to conclude the origins of the mentioned disasters in each continent.

Asia: Cumulative Effects of High Rate of Population Growth and Industrialization

The life quality in the river delta regions in Asia seems to be evidently connected to the rapid economic growth and low price labor emerged from the growing population in this continent. However, in this context, the pros and cons must be investigated.

The high population growth in Asia has changed a wide range of socio-economic equilibrium over the previous decades. An assessment of the global environment indicates that decisions made in China and India will be potentially one of the most important factors shaping the future global sustainability (Galli et al, 2012). In this section, a review over the current condition of major delta regions is provided for four high populated Asian countries; Bangladesh, china, India and Vietnam.

The change in the irrigation pattern is a potential cause of the declining life quality. In the early years of the 21th century, In Bangladesh, many hand pumps supplying the drink water has become dry because of the high level of groundwater diversion to the irrigation. As a consequence of this expansive irrigation activity, more that 20 million people are at the risk of water contaminated by arsenic. Furthermore, the concept of water as a good, not as a right, has become predominant by the growth of shrimp aquaculture for export (Crow and Sultana, 2002). Although it is very difficult to judge about the tradeoff between the increasing food production and the degradation of environment, but the negative impact of shrimp farms in the water salinity is a considerable issue that cast its shadow over the life standards in this region (Hossain and Dearing, 2013).

Although farming activities in Bangladesh were the major reason of challenges in the delta regions, other forms of human activities may also affect the environment in more catastrophic way. Yangtze River Delta (China) also experienced cropland loss due to the industrialization and population concentration which led to a considerable decrease in the grain production. This reduction has happened in two major phase between 1980–2005 and 2005–2010 with loss of about 80 and 66% of grain production respectively. The new industrialized zones have been established on the ground with high irrigation capability and high soil fertility. On the other hand, cropland recovery was applied extensively in regions with low irrigation capacity that caused the chronic food insecurity in Yangtze River Delta (Liu, 2015). There was also a similar pattern in the Pearl River delta, South of China, where the inflow of capital and industrial funds motivated a great number of rural laborers to leave agriculture in favor of factories in 1990's (Lin, 2001). By considering this pattern, the low quality soils must recover the vacancy of reach soil via increasing reliance on agrochemicals. Furthermore, the remaining potential fields are near the industrialized regions that may contaminated by industrial waste. Subsequently, there is an increasing anxiety about the accumulative sedimentation of harmful substances such as nitrates and heavy metals in agricultural soils and also the crops in China (Wong et al, 2002). This issue becomes more serious as there is evidence of correlation between the changed landscape pattern and water quality variables (Cl−, EC, NH3–N, and NO3–N) in these regions. As a case study, a research in the Yellow River delta reveals that the heavy metals and Arsenic were closely correlated and originated from common pollution sources (Bai et al, 2012). Despite the past, now it is very obligatory to establish an effective monitoring and management system to keep the minimum standards of a sustainable environment in the river systems of china damaged by human activities (Zhou, 2012).

The negative effects of the mentioned activities not only reduce the quality of life, but also may make the human life absolutely impossible. As one of the biggest river delta in the world, the Indus delta contains millions of people supplied by economic, environmental, and ecological benefits of the fresh water and the fertile soil. However, a chain of major dams in the upstream has disrupted the natural flow in a way that for most times no water is released to the Delta region; the condition that has brought the delta to the brink of death and annihilation (Memon, 2005). The condition is not better in the adjacent country; India. In the Cauvery delta, the quality of groundwater supplying both domestic and agricultural requirements has been investigated in various researches. In a case study, 55 % of samples gathered from shallow wells and wells in coastal parts could not fit for drinking regarding the World Health Organization and Bureau of Indian Standards. In another case, Mahanadi basin, a relatively dense concentration of heavy metals has been observed in the estuarine sediments (Sundaray et al, 2014).

Furthermore, a damaged environmental system can also affect the life quality in the form of short term difficulties, In the Mekong delta, in Vietnam, unhygienic water is known as the main reason of more than 90% of disease cases (Herbst et al, 2009). Furthermore, in this basin, hydropower development programs are recognized as the catastrophically threatening activities for fish biodiversity and food security (Ziv et al, 2012). In a larger scale, macro scale analysis indicates that Vietnamese landscape is subjected to challenging changes that can potentially defect the welfare in the national level; natural forest losses, shrinking biodiversity, greenhouse gas emissions, losses of paddy rice and other agricultural lands in the Red River Delta and the Mekong River delta. Theses disasters may be exacerbated by more frequent stormy floods (Rutten et al, 2014). A survey in the socio-economic data of 1999 shows that most of the poor live in the Red River Delta and the Mekong River Delta; However, the inequality is relatively lower than international average indicating the high spread of poverty among a large population (Minot et al, 2006). Changing the life style or agriculture techniques in this relatively large population group is not easy. However, a new approach to the agriculture system has become more essential because the current rice-maize system, as the traditional cropping system in the Vietnam major deltas, is reported not to be sustainable and its revenue margin is narrowing in the most part of the potential areas especially in the Red River delta (Le, 2014).

Although in the overwhelming majority of the existing researches, the direct effects of human activities have been mentioned as the main reason of the current uneven condition in the Asian river deltas, but some indirect effects of the global modernization must be considered to. For example, Mahanadi delta experienced a series of frequent drought and flood in the previous decades as a result of global climate change. Uneven distribution of precipitation (Heavy rainfall in May and September and severe drought in November and December) had a sizable negative effect on the crop yield. (Raychaudhuri et al, 2015)

Africa: the Silent Victim of Climate Change and Poverty

In the previous section, a review on the current researches about the life quality in Asian river deltas revealed that the direct effects of human activities, in both local and global scale, might be considered as the origin of the existing disasters. However, in this section, the results of the similar researches in African cases may be concluded as a relatively different outcome.

There are particular kinds of unexpected reasons that threaten the life quality of the residents in some African river deltas arose from the concentration of the large population and poverty. As a notably case, in the Okavango delta (Botswana), the general prevalence of HIV/AIDS has earnestly jammed fishing activities, and consequently the food security amongst the 29% of households (Ngwenya and Mosepele, 2007). In some other cases, there are economic attempts to increase the regional wealth to prevent from the social disasters; however it is not very simple to compare the two sides of this sword. In the Niger delta, following the dredging of an oil well access canal, undesirable changes happened in the water quality indexes such as the Ph., Dissolved Oxygen (DO), TDS and sulfate (Ohimain et al, 2008). Although it was expected that oil extraction would increase the regional wealth in the Niger delta, but the community development partnership applied by international oil companies for poverty reduction have not been effectively fruitful (Idemudia, 2009).

In addition to the mentioned issues, the impacts of climate change have been extensively argued in the recent literatures. These impacts seem to be more serious in Africa because of severe direct effects, high agricultural dependency, and limited capacity for the adaptation efforts (Collier, 2008). Kenya's Tana River Delta has experienced catastrophic challenges arose from the climate change. For thousand years, the agro-ecological production systems had been adopted to the flooding regime of the river; farming, fishery and livestock-rearing are the main resources of wealth that each of them blooms in certain periods of the flooding events. The local population had adopted its economic activity to this long term cycle. Despite this intellectual solution, the decrease in water availability affected each subcomponent of the production systems, which ruined the economic structure in the 1990s (Leauthaud et al, 2013). In terms of low capacity for adaptation efforts, the Senegal River delta is a notable case. Changes in temperature, relative humidity, precipitation and evaporation, in combination with the current water resources allocation pattern to the agriculture sector has accelerated the lands degradation and water salinity in this region. Consequently there is a strong need to new adapted approaches to assure sustainable food security in this area (Diack, 2015).

Even in more developed countries as like as Egypt, the negative effects of climate change are extensive; the Nile-Delta is subjected to shoreline erosion and see level rise both emerging from the climate change. Increasing temperature has accelerated the evaporation rate that exacerbated the decreasing water levels in the Nile River in combination with reduced precipitation. Consequently, rising urban population relying on groundwater put a higher pressure on the groundwater quality. This condition causes serious challenges to ensure the supply of drinking water and irrigation. (Mabrouk et al, 2013).

 Owing the outlined causes, the life quality in the African delta regions is threatened by accumulative effects of climate change, poverty and lack of adaptation capability. Although these reasons might be addressed in the Asian cases, but the level of engagement is considerably different between these two continents.

Conclusion

There is a long list of evidences indicates that access to the basic life requirements such as healthy food, fresh water, sanitation, work opportunities and other welfare factors is noticeably difficult in delta rivers in the developing countries. A review on the current literatures about the related challenges in Asian and African river deltas indicates that despite similarities, there are different sources for the major mentioned reasons. In Asia, beside the climatic reasons, the direct effects of industrialization, as a main factor, have exacerbated the negative effects of overcrowding. New forms of agriculture (in response to the growing global demand), extensive use of fertilizers to increase soil agricultural capacity, widespread damming projects, establishment of factories using low cost labor, and high amounts of urban and industrial disposals have been addressed in this context. In contrast, an overwhelming majority of the published researches about African delta regions outline the poverty and lack of adaptation capability as the factors that worsen the effects of climate change; a global phenomenon that less industrialized African nation have no major roles in, but they must compensate the undesirable impacts.

.

References

Bai, J., Xiao, R., Zhang, K., & Gao, H. (2012). Arsenic and heavy metal pollution in wetland soils from tidal freshwater and salt marshes before and after the flow-sediment regulation regime in the Yellow River Delta, China. Journal of Hydrology, 450, 244-253.

Collier, P., Conway, G., & Venables, T. (2008). Climate change and Africa. Oxford Review of Economic Policy, 24(2), 337-353.

Crow, B., & Sultana, F. (2002). Gender, class, and access to water: Three cases in a poor and crowded delta. Society &Natural Resources, 15(8), 709-724.

Diack, M., Diop, T., & Ndiaye, R. (2015). Restoration of Degraded Lands Affected by Salinization Process Under Climate Change Conditions: Impacts on Food Security in the Senegal River Valley. In Sustainable Intensification to Advance Food Security and Enhance Climate Resilience in Africa (pp. 275-288). Springer International Publishing.

Galli, A., Kitzes, J., Niccolucci, V., Wackernagel, M., Wada, Y., & Marchettini, N. (2012). Assessing the global environmental consequences of economic growth through the ecological footprint: a focus on China and India. Ecological Indicators, 17, 99-107.

 Herbst, S., Benedikter, S., Koester, U., Phan, N., Berger, C., Rechenburg, A., & Kistemann, T. (2009). Perceptions of water, sanitation and health: a case study from the Mekong Delta, Vietnam. Water Science and Technology, 60(3), 699.

Hossain, M. D., & Dearing, J. A. (2013). Recent trends of ecosystem services and human wellbeing in the Bangladesh delta. Working Paper of Assessing health, livelihoods, ecosystem services and poverty alleviation in populous deltas project, Southampton, GB, University of Southampton, 64pp. Retrieved from: http://eprints.soton.ac.uk/id/eprint/360583

Idemudia, U. (2009). Oil extraction and poverty reduction in the Niger Delta: a critical examination of partnership initiatives. Journal of Business Ethics, 90(1), 91-116.

Le, T. T. L. (2014). Assessment of the sustainability of the rice-maize cropping system in the Red River Delta of Vietnam and developing reduced tillage practices in rice-maize system in the area. Journal of Vietnamese Environment, 5(1), 1-7.

Leauthaud, C., Duvail, S., Hamerlynck, O., Paul, J. L., Cochet, H., Nyunja, J., ... & Grünberger, O. (2013). Floods and livelihoods: The impact of changing water resources on wetland agro-ecological production systems in the Tana River Delta, Kenya. Global Environmental Change, 23(1), 252-263.

Lin, G. C. (2001). Regional integration in South China: Processes and consequences in a local economy of the Pearl River Delta. Asian Perspective, 93-125.

Liu, P., & Lu, C. (2014, December). Natural and Human Impacts on Recent Development of Asian Large Rivers and Deltas. In AGU Fall Meeting Abstracts (Vol. 1, p. 0581).

Liu, G., Zhang, L., Zhang, Q., & Musyimi, Z. (2015). The response of grain production to changes in quantity and quality of cropland in Yangtze River Delta, China. Journal of the Science of Food and Agriculture, 95(3), 480-489.

Lowi, M. R. (1995). Rivers of conflict, rivers of peace. Journal of International Affairs, 49(1), 123.

Mabrouk, B., Arafa, S., Farahat, H., Badr, M., Gampe, D., & Ludwig, R. (2013, April). Impact of climate change on water and agriculture: Challenges and possible solutions for the Nile Delta. In EGU General Assembly Conference Abstracts (Vol. 15, p. 6293).

Memon, A. A. (2005, May). Devastation of the Indus river delta. In Proceedings of World Water & Environmental Resources Congress (pp. 14-19).

Minot, N., Baulch, B., & Epperecht, M. (2006). Poverty and inequality in Vietnam: Spatial patterns and geographic determinants (No. 148).

Ngwenya, B. N., & Mosepele, K. (2007). HIV/AIDS, artisanal fishing and food security in the Okavango Delta, Botswana. Physics and Chemistry of the Earth, Parts A/B/C, 32(15), 1339-1349.

Ohimain, E. I., Imoobe, T. O., & Bawo, D. D. (2008). Changes in water physico-chemical properties following the dredging of an oil well access canal in the Niger Delta. World J. Agric. Sci, 4(6), 752-758.

Raychaudhuri, M., Panda, D. K., Kumar, A., Srivastava, S. K., Anand, P. S. B., Raychaudhuri, S., & Kar, G. (2015). Impact of Climatic Variability on Crop Production in Mahanadi Delta Region of Odisha. In Climate Change Modelling, Planning and Policy for Agriculture (pp. 99-107). Springer India.

Rutten, M., van Dijk, M., van Rooij, W., & Hilderink, H. (2014). Land use dynamics, climate change, and food security in Vietnam: a global-to-local modeling approach. World Development, 59, 29-46.

Sundaray, S. K., Nayak, B. B., Lee, B. G., & Bhatta, D. (2014). Spatio-temporal dynamics of heavy metals in sediments of the river estuarine system: Mahanadi basin (India). Environmental earth sciences, 71(4), 1893-1909.

Susser, M. (1985). Epidemiology in the United States after World War II: the evolution of technique. Epidemiologic reviews, 7(1), 147-177.

Vetrimurugan, E., Elango, L., & Rajmohan, N. (2013). Sources of contaminants and groundwater quality in the coastal part of a river delta. International Journal of Environmental Science and Technology, 10(3), 473-486.

Wang, P., Wolf, S. A., Lassoie, J. P., & Dong, S. (2013). Compensation policy for displacement caused by dam construction in China: an institutional analysis. Geoforum, 48, 1-9.

Wong, S. C., Li, X. D., Zhang, G., Qi, S. H., & Min, Y. S. (2002). Heavy metals in agricultural soils of the Pearl River Delta, South China. Environmental Pollution, 119(1), 33-44.

Zhou, T., Wu, J., & Peng, S. (2012). Assessing the effects of landscape pattern on river water quality at multiple scales: a case study of the Dongjiang River watershed, China. Ecological Indicators, 23, 166-175.

Ziv, G., Baran, E., Nam, S., Rodríguez-Iturbe, I., & Levin, S. A. (2012). Trading-off fish biodiversity, food security, and hydropower in the Mekong River Basin. Proceedings of the National Academy of Sciences, 109(15), 5609-5614.